MW01101829

Glaciers and Icecaps

First English language edition published in 1998 by
New Holland (Publishers) Ltd
London - Cape Town - Sydney - Singapore

24 Nutford Place
London W1H 6DQ
United Kingdom

80 McKenzie Street
Cape Town 8001
South Africa

3/2 Aquatic Drive
Frenchs Forest, NSW 2086
Australia

First published in 1997 in The Netherlands as
Gletsjers en IJskappen by
Holding B. van Dobbenburgh bv, Nieuwkoop,
The Netherlands
Written by: Dr. Jaap J.M. van der Meer
Translated from the Dutch by: K.M.M. Hudson-Brazenall

Copyright © 1997 in text: Holding B. van Dobbenburgh bv,
Nieuwkoop, The Netherlands

Copyright © 1997 in photographs: individual
photographers and/or their agents as listed on page 2

Copyright © 1997: Holding B. van Dobbenburgh bv,
Nieuwkoop, The Netherlands

All rights reserved. No part of this publication may be
reproduced, stored in a retrieval system or transmitted, in
any form or by any means, electronic, mechanical, photo-
copying, recording or otherwise, without the prior written
permission of the publishers and copyright holders.

ISBN 1-85368-695-6

Editorial direction: D-Books International Publishing
Design: Meijster Design bv
Cover design: M.T. van Dobbenburgh

Reproduction by Unifoto International Pty, Ltd

Technical Production by D-Books International
Publishing/Agora United Graphic Services bv

Printed and bound in Spain by Egedsa, Sabadell

CONTENTS

PHOTO CREDITS

Dobbenburgh, B. van, 15, 29[t], 30, 36, 40, 55[t]; Dobbenburgh, M.T. van, 5[t], 16, 17, 18, 19, 20, 21, 22, 23, 24/25, 34, 35, 37, 38, 39, 42, 43, 47[t], 48, 52; Hazelhoff, F.F./Foto Natura, 4, 10[t], 11, 62; Helo, P., 79; King, V., 9, 47[b], 58, 59[b], 60, 61; Kluiters, J., 53, 64, 74; Lemmens, F./Foto Natura, 12, 55[b], 63, 67, 70, 74[b], 75; Meer, J.J.M. van der, 5[b], 6, 7, 26, 27, 30, 31, 32, 33, 44, 45[b], 51, 59[t], 68; Meinderts, W.A.M./Foto Natura, 13, 65, 71; Meurs, R. van/Foto Natura, 56/57, 66, 69, 77; _sterreichischer Alpenverein, 49; Sanders, A./Foto Natura, 28; Volgelenzang, L.,/Foto Natura 72/73, 76; Vugts, H.F., 14, 49; Weerink, W., 54[t]; Zwiters Verkeersbureau, 29[t], 41, 45[t], 46, 50;

r=right, l=left, t=top, b=bottom, c=centre

Introduction

In nature, water can be found in two states: it is either in motion or it is being stored. It is in motion as water vapour, rain or snow in the atmosphere or as running water in a river. In its other state, storage, we find water in seas, in lakes or held frozen as ice. Ice occurs on the earth's surface as glacier ice. If it is a small ice mass, we call it a glacier, if a large mass then we call it an icecap.

This book is about glaciers and icecaps, how they behave and grow or shrink. The various forms in which glaciers occur, and the shapes to be found on and in glaciers, as well as the role played by water in building and breaking up glaciers are all described in this book. Finally, there is a brief look at the traces left on the earth's surface by glaciers. But most of all, this is a book about what glaciers look like.

Glaciers are important to many people on earth. For some people, they are a source of fresh water for irrigation or for the production of electricity. To others, they are phenomena that make a lasting impression, or they are simply objects of natural beauty. They are of great importance to all of mankind because all the glaciers and icecaps on earth, from Antarctica to the mighty glaciers of the Arctic region, contain so much water that, if they were all to melt, the sea level would rise by at least 70 metres.

Glaciers and Icecaps

The simplest way to introduce glaciers, is to say they are the result of very heavy snowfall in the winter. It is naturally much more complicated than this, for there are many other places where enough snow falls and yet no glaciers form there because all the snow melts away the following summer. This shows that the creation of glaciers depends on at least two conditions: in the first place, snow must fall and in the second place, some of the snow must survive the following summer.

Apart from precipitation falling as snow, there are other ways in which sufficient snow can gather in one place; for instance, avalanches that cause the whole of the snow that has fallen over a wider area to gather in one spot. If such a collection area is a place where the sun does not shine, then it is relatively simple for a glacier to develop there. In other regions like Antarctica, very little snow falls each year, yet because it is so cold there almost no snow melts in the summer. The result is a gigantic icecap.

The icecap in the background feeds a wide valley glacier that ends in the sea. In the foreground various valley glaciers round a typical 'horn' in Greenland.

On the large Greenland glacier slowly more medial moraines appear, which divide the glacier into a number of parallel bands. In the background yet another small icecap.

The complex glacier world of the high mountains: snow-covered peaks and the bare ice of low-lying glacier tongues, a view of the icecap Eyafjallajökull, Iceland.

Icecaps are much simpler in structure: a snow-covered plateau and an irregular edge of bare ice. Drangajökull, NW Iceland.

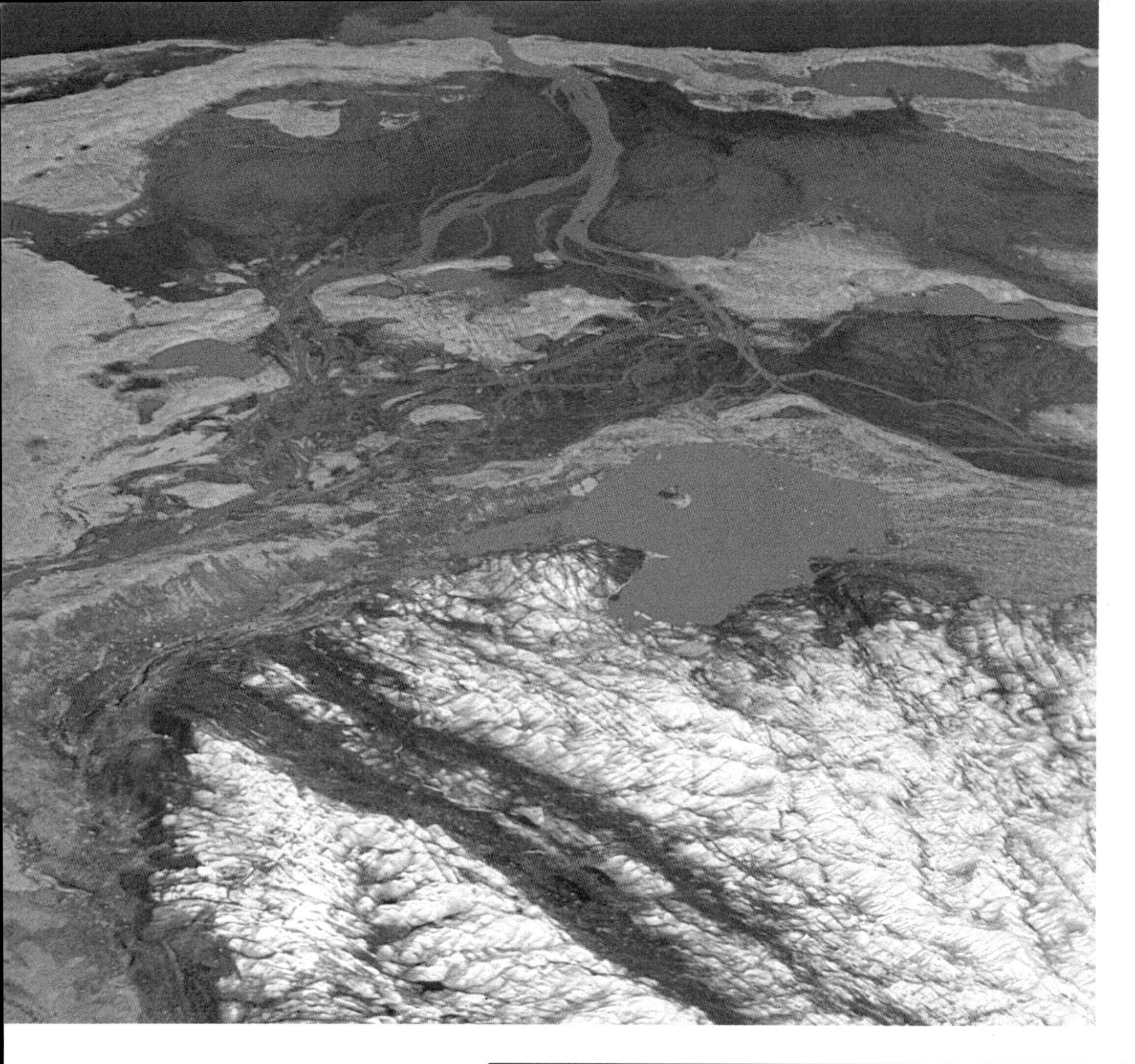

The valley glacier Serminguaq in SW Greenland, spread much further into the fiord 100 years ago. The gradual retreat of the ice has left behind a landscape with bow-shaped recessional moraines and meltwater channels.

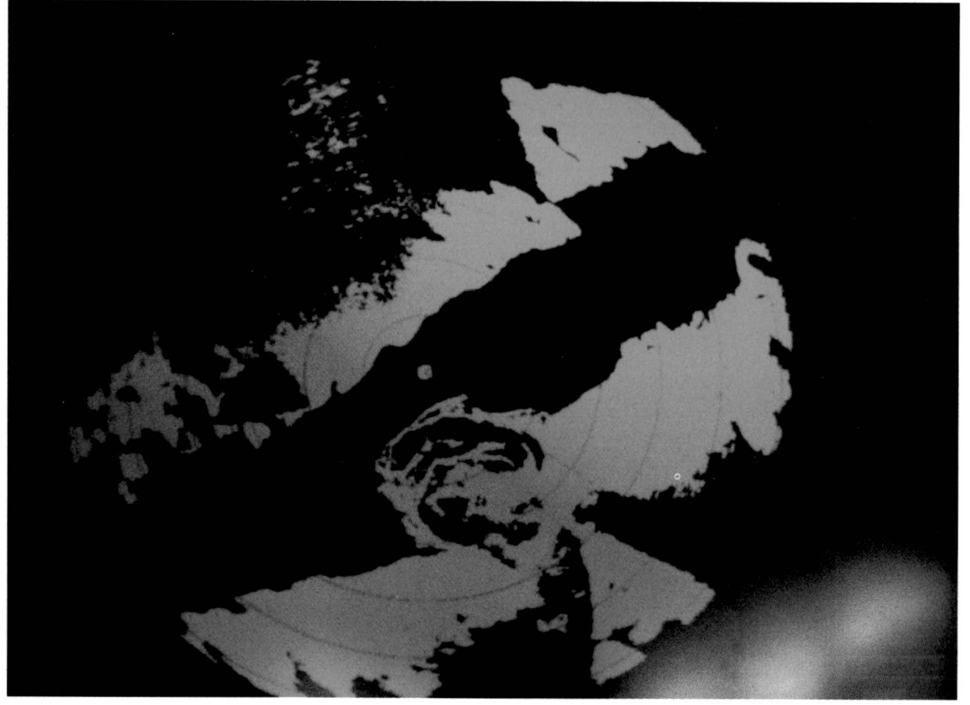

The radar image on board a ship shows the straight sides of the fiord. Serminguaq is visible as a round outwash, whereby the parallel moraines can be clearly seen in the fiord. It is easy to see that the glacier nearly dammed up the fiord.

It is plain to everyone that snow is not the same as ice, and that something has to happen to the snow that has fallen before it can be called ice. It does take a number of years before the transformation from snow to ice is completed. What happens is that the light snow layer has to be changed into a solid layer of ice, which is, in part, the result of compression of the snow by the layer of new snow lying above it. At the same time, the snow on top melts slightly and the meltwater thus formed trickles down through the layer. Deeper into the layer of snow, the snow is much colder and there the meltwater freezes in the empty spaces between the snow crystals. After a

The hills of SW Greenland disappear under the edge of the land ice: the horizon only shows the smooth surface of the ice. In the foreground we can see the typical image of a chaotic meltwater river.

few years the result is first a 'névé' layer (granular snow) and thereafter a thick layer of ice in which air can only be found in bubbles. Since these transformations do not always take place in the same way or at the same speed, we are often able to observe differences in colour and clarity of the glacier ice that has formed. Sometimes it is white, with lots of air bubbles, or it can be as clear as glass. In a steep section of the glacier or in the crevasses that occur there, various coloured layers of ice can be observed. In particular, when the sun shines on it, many different shades of blue can be seen.

If we want to know how a glacier works, it is best to look at the annual cycle of a small glacier. A look at a small glacier is sufficient, because in principle there is no difference between a large and a small glacier. In the winter, snow falls over the entire area of a small glacier, increasing the mass, but in the following spring when this snow begins to melt, it not only melts on the snout of the glacier but higher up as well. As the air at the snout is warmer than higher up on the glacier, the snow at the snout will melt faster. By the end of the summer the snow at the snout will have disappeared completely, leaving the ice exposed. The line that forms the boundary between the areas still covered with snow and the snow-free section of the glacier is called the snowline. In the summer, the snowline can be seen on all glaciers and then it is clear that it is not a straight horizontal line. Glaciers are not flat tables, so the snowline is irregular, as can be seen in many of the photographs shown here.

Glaciers in the temperate zones lose the greatest proportion of their snow and of their ice too, through melting in the summer. This is evident from all meltwater rivers that rise under glaciers. However, glaciers that discharge into the sea or a lake also lose part of their snow and ice mass through the calving of icebergs. If our small glacier were to discharge into the sea, it would only produce small icebergs. The large ice masses of Antarctica, on the other hand, when they reach the deep ocean, can calve icebergs of many hundreds of square kilometres, which then drift off into the ocean.

If, for a couple of years, more snow falls than normally survives the summer on our small glacier, then the glacier increases its total mass. This can be best observed at the snout of the glacier which advances. We say that the glacier 'grows'. If there is less, rather than more snow left over, the mass of the glacier reduces and the snout retreats. Such changes in the position of the glacier's snout are quite normal over a period of time and throughout the history of the earth. If we look at the long term, we can see that glaciers covered vast areas of the northern hemisphere continents during the ice ages. 20,000 years ago in North America, there was an icecap that was larger than present day Antarctica. However, we can also see that glaciers change in size over a shorter time scale. If we look at the postcards sent by our grandparents from Switzerland, the difference in the same glaciers is plain to see, since all glaciers in the world were greater in volume last century than they are nowadays. That is not the result of the greenhouse effect but of natural climatic variation; although, if the greenhouse effect continues, then it will have dramatic consequences for the glaciers on earth.

The snout of a glacier can only advance if the ice itself moves, and that is precisely what happens. Very slowly, indiscernibly slowly, the ice creeps downhill from the regions where the snow has survived the summer to the regions where it has not. Gravity, of course, lends a hand. If our small model glacier is to remain in the same place or rather remain in balance, then every year, all the snow and ice that melts has to be replaced by ice that flows from higher up. Although this process is not visible to the casual observer, all glaciers are very slowly flowing downhill, rather like very thick syrup. The glacier that holds the world speed record is the Jacobhavn's Isbrae in the west of Greenland, which maintains a constant speed of nine kilometres per year; that is slightly faster than one metre per hour. From time to time, other glaciers exceed the speed of Jacobhavn's Isbrae, but only for a few months at a time; then their speed drops back to their normal sluggish pace. Glacier surges like this occur in many places on earth.

Schematic drawing of a glacier. In the dark blue section accumulation occurs over a whole year. The ice thus formed, slides slowly down to lower-lying regions. The cross section shows where most debris is carried in the glacier. At the snout of the glacier this debris forms morainic walls.

There are many types of glaciers and icecaps, which are normally categorised according to size, with ice sheets being the largest, icecaps being somewhat smaller, glaciers next down in size and névé fields the smallest. There are only two real ice sheets at present on earth, Greenland and Antarctica. These two ice sheets are so large, that they completely cover the underlying earth's crust. This cover is so complete that we cannot even see whether there are mountains like the Alps or plains like those of north-western Europe lying under them. They are of such an immense size, and this is true of all larger icecaps, that the flow of the ice is not influenced by the underlying surface. The only thing that we are able to measure on the surface is that the ice does flow from the highest point directly to the edge. Smaller ice masses like the familiar valley glaciers are, in contrast, highly influenced by the topography of the surface. A glacier in the Alps, the Himalayas or the Rocky Mountains can only flow downhill in its own valley, for such a glacier cannot cross over to the next valley. This is not the case with icecaps: glacial flows cross entire mountain regions as if they did not exist.

A glacier that ends in the sea or in a lake, like the Perito Moren Glacier in Argentina, loses a great deal of its mass through calving, the formation of icebergs, and less through melting.

In many regions with glaciers we can see small glaciers in isolated cirques in the mountains. Sometimes these cirques look like comfortable

armchairs in which the glacier leans over backwards. Such small glaciers are known as 'cirque' glaciers, after the hollow or cirque in which they lie. There is only one kind of ice mass that is smaller and that is the névé field. The latter is not much more than a snow patch in which the snow has scarcely been turned to ice (névé). Usually there is not much movement in such ice patches.

There are also other different types of glaciers, such as the aprons of glaciers on and around isolated volcanoes. An example of this can be seen in part of the Rocky Mountains in North America and in the Andes in South America. The symmetrical shape of the volcanoes often allows the apron of glaciers to form a neat circle around the top.

The crests of the high mountains in Nepal are covered with snow. That is why it is not easy to see where the crevasses lie.

If not all the newly fallen snow melts away, then it ultimately turns into glacier ice. As the circumstances on each slope are different, this transformation is much easier in one place than in another. This is a picture from India.

The highest peaks of high mountains are so steep that the snow never lies for long enough to form glaciers. However, as fresh snowfalls do occur regularly, they often look as if they have been covered with icing sugar.

Another special kind of glacier is found below very steep slopes. A glacier that develops above a very steep slope, creeps steadily down towards it. Once the glacier arrives at the edge of the precipice, large blocks of ice break off and fall down. At the bottom of the precipice the blocks of ice freeze back together again and the glacier flows further as if nothing had happened. Regenerated glaciers can be found in all high mountain regions on earth.

In the Arctic region, glaciers which began life as part of an icecap are found. The ice in the icecap moves more or less independently of the underlying relief, but we often see that, in a few places, it moves in the form of ice tongues. Such glaciers are often long and thin and lie in shallow valleys, and since these are very steep, the glaciers are very broken up and accidented. Sometimes, such glaciers can span a height difference of more than 1,000 metres. When these glaciers open out onto a flat valley floor, they often display a different shape, namely, a fan shape. Such glacier lobes are called 'piedmont' glaciers, after the Italian region of the Piedmont. Very large glacier lobes with this typical shape lay there during the Ice Ages; this can be seen in the shape of Lake Garda, for instance. Piedmont glaciers, moreover, often display beautiful ice waves or 'ogives', which will be explained later.

Above the crest, sticky snow is building up a dangerous overhanging edge of snow. The grooves in the steep slopes of the Illigami in Bolivia show how avalanches of snow feed the glacier lying below.

The vertical slopes of the Fitzroy group in Argentina look bare once all the snow has fallen onto the lower-lying glaciers. Climbers, however, must still reckon on walls covered in ice.

Structures

Up to now we have looked at the shape and size of glaciers as a whole, but we can also look at the components of glaciers and of their surfaces. When we do so, it appears that there is great variation in shapes and phenomena on the glacier surface, some of these forms relating to water that flows over and within the glacier. We will look at that later in closer detail. In this section on structures we will look at the forms of glacier ice itself.
In many places in the world, we see that there are mountains and sections

of mountain protruding through the ice. Apparently, the ice in these places is not thick enough to obscure the underlying land forms entirely. Such isolated sections of rock are known by the Greenland name 'nunatak'. These are not simply lumps of rock, for in regions with large volumes of ice, these are the only places that harbour plants and animals. They are the reason that we are able to find life in the most barren places in the world. In the Arctic region, nunataks are favoured by birds as nesting places; they offer protection against predators like the Arctic fox. In Antarctica, very large nunataks can be found: whole mountains with valleys lie in the

The horizontal névé fields of a glacier in Canada are surrounded by steep slopes. The snow surface is only broken here and there by sharp rims that betray the presence of crevasses. After passing the ice fall, the glacier once again encounters a flatter piece of ground.

△The Nisquilly Glacier in Canada lies on a mountain with such steep sides that the glaciers seem to consist almost entirely of ice falls. On the steepest sections, blocks of ice break off and tumble down in a single fall. Then they melt back together and the glacier continues downwards.

△▷The form of the underlying mountain determines the final shape of the glacier. Yet ice that splits apart high up on the slope joins back together down in the valley. The crevasses show where the ice traverses steep slopes (transverse crevasses) or where it fans out (radial crevasses), Mt Baker USA.

▷Virgin snow reaches as far as the eye can see. Aiguille du Midi, Mont Blanc massif, France.

middle of the bare expanse of ice. They are considered to be oases because, like oases in the desert, they are entirely surrounded by a hostile environment. These are the only places in Antarctica where flowing water can sometimes be found.

In the previous chapter we saw that glaciers grow because a layer of snow remains each year. As no two years are the same, the layers of snow are slightly different from year to year. This enables us, even after thousands of years, to observe layers in the glacier ice. In ice cores taken from the Greenland Ice Sheet, some 1,500 layers could be counted. The fact that we can see layers, is not only due to characteristics of the ice itself, but is also caused by the presence of other material in the glacier. There are, for example, very thin layers of ash derived from volcanic eruptions or dust layers from dust storms in the great deserts. The ash clouds resulting from major volcanic eruptions travel around the world in a very short time, after which the ash begins to fall to the surface. In the icecaps of Greenland and Antarctica, these layers of ash have been carefully preserved and many are known from ice cores. As long as the glacier flows smoothly over a flat surface, these layers are not disturbed and they are even still recognisable at the snout of the glacier. It was possible to trace some of the great volcanic eruptions, i.e. Krakatau and Tambora in the 19th century in ice cores from Greenland. This research also revealed prehistoric volcanic eruptions. Traces have also been found of an eruption from the time of ancient

In the foreground we can see the nursery area, where the glacial ice is formed. Through the veil of cloud we can see how they form, at a much lower level. The Mer de Glace (France), a veritable 'highway' of glaciers.

Greece, which was apparently the Santorini volcano, the pretext for the story of Atlantis. In many instances, however, glaciers do not flow over level ground. Instead, in every high mountainous region we can see that glaciers move down over steep slopes. Although ice can flow like thick syrup, such steep slopes are too much even for glacier 'syrup'. The ice simply cannot follow the slope fast enough and it therefore breaks into blocks. When this happens, we say that the glacier has formed an ice fall, just like running water forms a waterfall. When a glacier has crumbled into many blocks, then an ice fall consists solely of independent pinnacles of ice, known by the French name 'serac'. This produces spectacular images but for climbers, such ice falls, with their seracs and crevasses, can be insurmountable.

The flow of glacier ice is not interrupted by the seasons; the ice continues to flow throughout the year, but higher temperatures mean that it flows faster in the summer. This means that more ice passes through an ice fall in the summer than in the winter. At the base of the ice fall, where the ice freezes back together, this difference in speed can be easily observed. Here, large semicircular-shaped ridges of ice develop, called ogives. The increased volume of ice in the summer forms a ridge that can be several tens of metres high. It takes a few years before an ogive or 'ice wave' has melted away enough so that it is no longer recognisable. For this reason, we can often observe under the ice fall of many glaciers, such as the piedmont glaciers mentioned above, a similar pattern of parallel ice arches. Dozens of arches can be seen gradually reducing in height. High on the glacier, approximately where the glacier touches the mountainside, something strange occurs. On the one side, it is so cold that the glacier is frozen solid to the wall, but on the other side the glacier is always trying to flow

Glaciers are essentially different from one another. While here various glaciers lie alongside each other, one consists only of closely packed crevasses, whereas the other shows an unbroken snow surface. Glacier des Bossons with the Glacier de Taconnaz behind it, in the Mont Blanc region, France.

If the air temperature is low enough, then glaciers can also grow through water vapour from the clouds freezing on the surface. This happens here on the Aiguille du Midi in France.

downhill. This produces so much tension in the ice that it splits. This tension is omnipresent, so a glacial split can always be found, parallel to the mountainside. Mountain climbers know this deep, wide crevasse as the 'Bergschrund'. It is one of the many sorts of glacier splits or crevasses, but fortunately for climbers it is almost always visible and is only hidden from view after fresh snowfall. There are many places on glaciers where so much tension is created that the ice splits with a bang. This is the case, for example, where the ice flows along the mountainside. The friction along the mountainside slows the speed of the glacier, but towards the middle of the glacier, this friction does not occur and the glacier can flow there at its normal speed. The difference in speed results in the formation of wide, open crevasses at right angles to the edge. Crevasses also develop in this way, in places where a glacier passes over a rocky threshold, or where the glacier valley bends. Since there will always be mountainsides, rocky thresholds and bends, new crevasses will continuously develop at those points. Through the perpetual movement of the glacier, crevasses will always be moved away from the place where they develop. In other places, given that the pressure that formed them is no longer present, they will be slowly squeezed closed again. Even then, they can still be followed for some distance, until they close completely and only a thin line in the ice remains.

◁△Glaciers disappear through melting. Most meltwater only becomes visible once it appears from under the snout of the glacier. In the close-up taken halfway down the Glacier d'Argentière (France), meltwater is temporarily visible as it cuts the corner.

These thin lines remain recognisable in the ice until the glacier melts or breaks up because of the difference in the crystalline structure of the unsplit ice.

In the large level cirque, thick ice has formed, which slowly moves down the valley. This is where the fight begins between melting and accumulation, depending on warmth and precipitation. Val d'Anniviers, Switzerland.

In high mountains, we see steep mountain walls and tops rising above the ice. Closer examination reveals that pieces of rock are continually breaking off these bare sides. The action of frost causes the rock fragments to break off, and gravity and further rock falls carry them down onto the ice. Sometimes entire mountainsides collapse, resulting in large sections of the glacier surface being covered in debris. This is called 'supraglacial' debris or 'debris on the ice', in contrast to 'debris in the ice'. It is clear that such

debris only occurs when there are mountains protruding beyond the ice. A search for supraglacial debris in the Greenland or Antarctic ice sheets would be a fruitless exercise. The continuous flow of the ice ensures that the debris is carried slowly downhill on valley glaciers. This is most clearly demonstrated around nunataks, where the debris lands on the ice on all sides. It is then carried along with the moving ice as the ice flows past the nunatak. Where the moving ice meets on the downstream side of the nunatak, we see a trail of debris from the point of convergence, called a medial moraine that runs over the ice down to the snout of the glacier. The

▷Whether or not glaciers form, depends on the shape and position of the mountains. This explains why one slope is bare and the other is covered with a glacier. Zinal, Switzerland.

clearest medial moraines to see, are similar debris trails in combined glaciers, where a number of valley glaciers meet and continue as a single, large, complex, main glacier. Each of the original glaciers can be traced on the basis of the medial moraines that form the boundary with the neighbouring glaciers.

Glaciers above lakes are often regarded as dangerous. Blocks of ice that plunge from the steep glacial slopes can cause a dangerous flood wave.

▷Mountain trips, walking or climbing, usually go hand in hand with glaciers. In the Alps this is the reason why many Alpine huts stand very close to glaciers, like the Tracuit Hut here in Switzerland.

The high mountains show water in all its forms, as vapour in the air, as solid ice and liquid in the river. The vapour from the air forms the precipitation from which the glacier grows. If the ice melts, the water flows away to be evaporated into water vapour again, completing the water cycle. Susten Pass, Switzerland, on the left, the Stein Glacier and on the right, the Steinlimi Glacier.

High on the mountain, ice is formed from snow. Only ice lies one level lower, splitting and melting. Even lower, ice with debris is found.

Sometimes medial moraines look like highways with parallel lanes. This phenomenon can easily be seen in a number of photographs in this book. Whilst the debris is being transported by the glacier, it plays a special role.

In early summer the meltwater tunnels under the ice are still forming and the glacial gate is still mainly blocked. Turtmann Glacier, Switzerland.

A fine example of a glacial table stands on the Turtmann Glacier. The flat rock protects the ice on which it stands. Whilst the surrounding ice melts, the rock slowly acquires a pedestal.

The front of the Turtmann Glacier is unstable around the meltwater gate. A large pillar of ice leans threateningly.

Crevasses are typical of glaciers. In the foreground they together form an ice fall, which is characteristic of steep glaciers. Below the mountainsides in the background, the ice is tearing itself loose from the rock walls creating the so-called 'Bergschrund'. This phenomenon is found on all glaciers. Jungfrau, Switzerland.

During winter in the Jungfrau, the fresh snow hides the Bergschrund, but that does not mean it is not there.

Viewed from among the pines, the largest glacier in the Alps, the Aletsch Glacier in Switzerland, demonstrates its parallel medial moraines.

Even in the Alps, glaciers can reach a considerable depth. Ice more than 700 metres thick has been measured on the Konkordiaplatz (Aletsch Glacier).

△◁*A real summer picture of a glacier: the snow has melted, the crevasses are wide open and a round meltwater lake has drained. Gorner Glacier, Switzerland.*

△▷*The scene is quite different in early summer when the snow still covers the ice, whilst the glacier movement again pulls everything apart.*

▽◁*Glaciers carry debris and build side moraines from this material. Here, two glaciers from the Monte Rosa massif have formed a triangle of lateral moraines.*

▽▷*In the lateral wall of the Tsjidiore Nouve Glacier in Switzerland, debris is visible as thick bands.*

Small, dark pieces of debris collect enough warmth in the daytime to melt a hollow in the surface of the ice. Large stones and blocks are much too dense to allow the sun's warmth to reach the underlying ice, therefore they protect the ice on which they lie rather than melting a hole in it. Whilst the neighbouring clean ice melts away and steadily becomes lower, the debris-covered ice does not melt and it gradually comes to stand above the rest. For this reason we see ridges and piles of debris on so many glaciers. These look as if they are composed entirely of sand and gravel but, in fact, they are no more than a thin layer of debris on a core of ice.

Flat blocks of stone have an extremely remarkable way of protecting the underlying ice because, whilst the ice is protected against being melted, it is also carrying the block of stone. If enough clean ice has melted away around the block, then we observe a block resting on a pedestal of ice, which is called a glacier table. Sometimes we see whole processions of glacier tables. Of course, other objects can form tables on the ice; they only have to be large enough to prevent the sun's warmth reaching the ice, and stable enough to remain lying on the ice. This is the case, for instance, with sods of grass brought down onto the surface of a glacier by landslides.

◁At the edge of the Greenland Icecap the land ice fans out. Where this does not flow fast enough it tears apart in a regular pattern.

◁The smallest imaginable glacier. The snow lies for long enough to transform to ice, but there is no movement. During an ice age the glaciers that finally form icecaps and cover the land are formed here. Augsttelli, Switzerland.

▽◁Where the snout of a glacier fans out on a flat valley floor, then longitudinal crevasses occur on the edge. It looks like the Turtmann Glacier has fingers. Switzerland.

▽Higher up on the glacier the crevasses are squeezed back together before the walls of the crevasse can melt away too much.

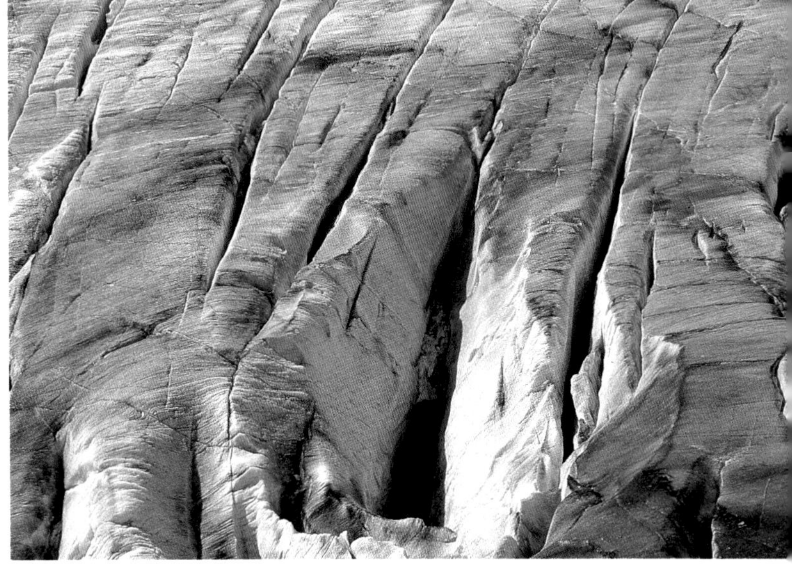

△▷*Various surface phenomena on the Grenz and Mutt Glaciers in Switzerland. Both photographs show a similar pattern of crevasses. Smaller crevasses and two parallel meltwater streams can also be seen on the left hand photograph, whilst the right hand photograph shows some fine snowdrifts.*

This brings us to life on glacier ice. In contrast to what one might expect, glaciers and snowfields are not entirely without life. For example, it is quite common to find red patches on the snow or ice in the summer. These are algae that live in the uppermost layers of the ice and then serve as food for other organisms like glacier fleas. These are not real fleas, but a small type of springtail that only lives on the ice. We can also find higher order plants on glaciers. It is, therefore, not unusual to see cushions of moss on the ice in the vicinity of bird nesting rocks (nunataks). The moss depends on the nutrients in the birds' droppings, and all the while, it is being moved along by the glacier.

The same happens if glaciers, covered in a thick layer of debris, end up close to sea level. Since these glaciers only melt slowly because of the covering layer of debris, and they descend into warmer regions, it is possible to find trees growing on them. This happens, for instance, on the Malaspina Glacier in Alaska and on the Casa Panque Glacier in Chile, where thick woods grow. The demise of the woods comes when they reach the snout of the glacier, where the trees fall over one by one and die.

In the coastal areas of the Arctic and Antarctic regions, we find, both on and under the ice, larger animals such as various species of seals. They are often seen best when they are sunbathing on a cold icefloe or iceberg. The only creature they have to watch out for is a polar bear. This formidable

The debris under a glacier leaves clear traces behind: stones sometimes cause half-moon gouges in the rock surface, whilst the fine material in the ice polishes the rock around it. Tacan, Søndre Strømfjord, Greenland.

The debris in the ice is sometimes able to make its way upwards, as seen here in a slide plane on the Serminguaq Glacier in Southwest Greenland.

Generally, glacier the ice is clean and contains no debris. Only differences in density cause differences in structure and colour. The section of the Russell Glacier in Greenland shown here is at least ten metres high.

hunter only occurs in the Arctic region, especially around the coasts and on the pack ice, but there have been reports of unexpected sightings of polar bears in the summer on the Greenland Ice Sheet and on the Arctic islands of Canada.

In the never-ending cycle from water vapour to snow and ice and then back to meltwater, we see water bubbling out from below the snow-covered glacier edge of the Glacier de Moiry in Switzerland.

If glaciers end up in the sea or in lakes, they begin to calve. The front becomes vertical and therefore unstable. The deeper the water at the coast and the larger the glacier, the larger the iceberg that will break off the glacier. The largest icebergs occur in Antarctica, where they may reach a size of hundreds of square kilometres. Although it is a majestic sight to see a huge rugged mass of ice floating in the sea, we have to

Glaciers slowly but surely break down the mountains they stand on. Here, the Glacier d'Argentière in France has formed a deep trough.

The troughs cut out by glaciers can be very deep. Even when the ice has disappeared, the shape of the trough shows that it has been shaped by ice and not by a river. The Glacier d'Argentière lies here in its own trough.

Snow, ice and meltwater. The proportion of these depends on the position in the landscape. Thus we can observe large and small glaciers alongside each other, as here in Saas Fee in Switzerland.

Fed by glacial water, the lower regions of the mountain range are green and attractive. Switzerland.

▷The size of glaciers and their associated phenomena is always difficult to assess. Is this a small meltwater stream that you could just step over or a raging torrent? Rhône Glacier, Switzerland.

bear in mind that we actually only see the tip of the iceberg. Only about 1/9 of the ice sticks out above water.

This means that for every metre above water there are another 8 metres below: an iceberg that seems to be 20 metres high, is in fact 180 metres high. Their immense depth under water can also cause them to run aground easily on the sea bottom, posing a threat to underwater cables and pipelines that have been laid there. Icebergs are fortunately no longer a real danger to shipping; thanks to radar they can also be visualised and avoided even in thick fog. Oil drilling platforms, on the other hand, are not mobile and a large iceberg is not easy to tow away. Off the coast of Labrador, oil platforms are moved if an iceberg seems to be on a collision course. This is safer than trying to tow or push the iceberg away. It is nevertheless not so long ago that icebergs did pose a great danger to shipping. The recent attempts to raise a section of the Titanic are a stark reminder.

▽▷It seems as if glaciers lie immobile in one place. However, the change in the snowline in the course of the summer and the crevasses show that it is a dynamic system. Val d'Anniviers, Switzerland.

One of the most frequently photographed places in the Alps, the Rhône Glacier at the Belvédère. This is the spot where the glacier moves from a relatively level slope to a steep section. The large, lateral moraine and the bare, light-coloured rocks show how large the glacier was 100 years ago.

It is only when humans hike over the glacier, as here on the Rhône Glacier, that we can get a sense of the size and scale of the crevasses, and of the scale of the landscape.

Yet, there have been serious studies into the possibility of towing icebergs from Antarctica to, for example, Saudi Arabia. In this instance, the idea was to place auxiliary engines against large, flat table icebergs, cover them with foil and tow the whole lot slowly north. This would have provided a large supply of fresh water of excellent quality. Technically, it all seemed to be perfectly feasible, but it was suggested that an iceberg of several square kilometres would have an adverse effect on the local climate, not to mention the freshening of the sea water around it. For the present, it also

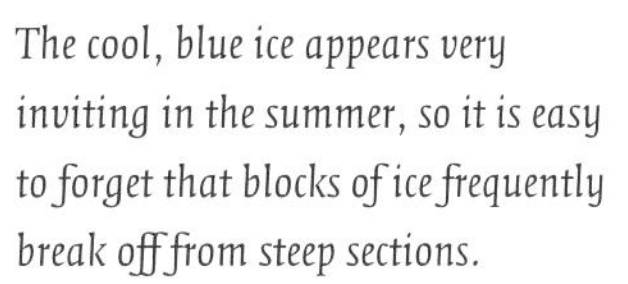

The cool, blue ice appears very inviting in the summer, so it is easy to forget that blocks of ice frequently break off from steep sections.

The symbol of a glaciated world, the Matterhorn in Switzerland. Surrounded on all sides by gnawing glaciers, all that is left is a sharp-pointed peak, or 'horn'.

remains cheaper to desalinate sea water; therefore there is not much chance of encountering an iceberg on tow whilst on a cruise near the equator.

Another attempt to use icebergs concerned an Allied attempt, during the Second World War, to build artificial icebergs from a mixture of water and sawdust. Such a mixture produces ice blocks that are stronger than pure ice. By building a hollow and insulated iceberg from this material and fitting it with engines, the Allies hoped to station a permanent aircraft carrier in the northern Atlantic Ocean. Just like a real iceberg, these artificial icebergs would be unsinkable; even military action could not affect them that much. A trial model, about 12 metres long, was actually built in

The Finsteraar Horn above the Obere Grindelwald Glacier in Switzerland. Here the ice has scraped out a deep, steep-sided trough.

The willow wood that has grown in the area abandoned by the Grindelwald Glacier is evidence of its greater size in the 19th century.

Just like in many other glaciers, a tunnel is cut into the Obere Grindelwald Glacier each year, allowing everyone to see inside.

Lake Patricia in Canada, but the war ended before a real ship, the 'Habbakuk', could be built. It would have been the largest ship in the world, with a length of more than 600 metres.

The importance of glaciers to people was mentioned in the introduction, including their use as a source of water for irrigation. In Wallis, Switzerland, meltwater is still a much used source of water for the irrigation of pastures and fields, and this is certainly not the only region in the world to do so. We must not forget that 95% of the fresh water on earth occurs as ice, mostly in Antarctica. In the Andes in South America, ice

It seems to be quite dark inside and the fantastic colour of the ice of the Rhône Glacier is only visible when one looks back towards the entrance.

Nature also makes tunnels in the ice. Here we can see the remains of a tunnel in an iceberg, Alaska.

Glaciers often follow a stepped surface, where it flows in waves over the steepest sections, like here near Grimentz in Switzerland.

Deep crevasses reveal the build-up of the Stein Glacier in Switzerland. Clouds add an element of mystery.

Between 1650 and 1900, glaciers all over the world were larger than at present: nowadays we talk of the Little Ice Age. The print on the left shows that the Pastern Glacier in Austria was much larger at the end of the 19th century than it is now. The cirque glaciers in the background have also lost much of their former size.

blocks from glaciers are packed in straw and then brought down to be used for chilling. This seems laborious and primitive, but from the 19th century onwards there was a regular trade in glacier ice, which lasted right up until after the Second World War. Before the middle of the 19th century, there was a demand in England for ice, for instance, for preserving fish. Ice could not be produced artificially but could be obtained from Norway; so trade began in blocks of glacier ice, which were cut directly from glaciers or fished up out of glacier lakes. The ice was covered with sawdust and transported in ships to England and traded there. In North America, a similar trade existed, not only in glacier ice but in ice sawn out of frozen lakes. Nowadays, glacier ice finds its way into whisky glasses in more luxurious surroundings, for the tinkle of glacier ice on the glass and the purity of the water apparently makes it worthwhile using this expensive commodity.

Nowadays, the major use of glaciers is for the production of electricity. Glaciers either drain directly into reservoirs, or the meltwater is collected immediately below the glacier and piped directly to a generating station. In addition, glaciers are being used for summer skiing much more frequently. Although this is undoubtedly a pleasant way of spending leisure time, it changes the surface of the glacier and increases the rate of melting. In an age in which many glaciers are visibly retreating rapidly, this cannot be recommended.

Glaciers in some regions are important as sources of water for power generation and for irrigation. In most regions they are also important as tourist attractions. Kleine Scheidegg and Eiger, Switzerland.

Water and ice

As indicated in this chapter, glaciers increase through snowfall and reduce through calving and melting. However, given that calving only occurs when glaciers end up in the sea or in lakes, melting is the most common form of reduction.

Most glaciers melt over their entire surface area, from the highest point to the snout. The situation is different with icecaps; in the highest regions of Antarctica melting never occurs. Here, the air above the ice is so cold that the temperature, even on the warmest day of the year, is still below zero.

In this chapter, we will examine what happens when glaciers melt and we will also take a look at the route taken by the meltwater to the glacier's snout. In the winter months, there is usually very little meltwater, therefore we will begin in the spring. It is then that the winter snow cover begins to melt, first on the snout, and later, higher up on the glacier. At this time, snow begins to melt on the upper side, but the meltwater cannot really flow away over the snow-covered surface. Instead, it sinks down into the snow and the snow cover slowly becomes saturated with water. On the glacier we find a thick layer of slush: this is a mixture of snow and water. The slush finally becomes so heavy that it begins to slide over the glacier ice. The friction generated by this causes more snow to melt. In the meantime, a slush avalanche has been created. This avalanche finally slides off the glacier and disappears in the foreland. It is also possible for all the snow in the slush to melt before it reaches the glacier's snout, forming a river on the ice itself. Meanwhile, the snow higher up on the glacier has also started to melt, but because melting at this level is a much slower process than lower down and the slope of the glacier is much steeper, much less slush is formed. Slush develops most frequently on the Greenland Ice Sheet and on the smaller icecaps of the Arctic islands.

On a flat section of the Turtmann Glacier in Switzerland the meltwater has developed an attractive pattern of streamlets and hummocks.

When all the winter snow has melted on the lower sections of the glacier, then the ice too begins to melt. The meltwater from the ice flows over the surface for it cannot sink into the solid ice. If more meltwater gathers, small rivulets soon develop. Small rivulets combine to form a stream and when they join up, we finally see a real river running over the ice. As the ice, over which many of these rivers flow, is so smooth, such meltwater rivers generally form symmetrical bends or meanders. Only small streams will form on most glaciers, but on icecaps, gigantic river systems can develop that are almost impassable. There is a belt of meltwater rivers and lakes that is about 100 kilometres across on the western side of the Greenland Ice Sheet. Since these rivers and lakes are all linked to each other, hikers crossing the icecap by foot frequently get into difficulties. A meltwater river

▽▷All glaciers end up melting. Here a glacier fills a small lake.

on the ice flows very quickly, the bottom is very smooth and the water is very cold. The combination of these three characteristics makes it very difficult to cross such a river, even if it is only a couple of metres wide.

The glacier surface is almost never flat and smooth, but is generally rough. This is why lakes can form on glaciers. Lakes of several tens of metres in width can be found on the Gorner Glacier in Switzerland, and also on the Meren Glacier in Irian Jaya. However, lakes of several kilometres across and several metres deep can be found around the fringes of the Greenland Ice Sheet. If one has the good fortune to fly over the Greenland Ice Sheet, then these lakes are visible as clear blue patches that stand out in stark contrast to the white ice. They are so blue that they are more reminiscent of warm Pacific beaches than the chilling cold of glacial ice. Most depressions in the glacial surface are caused by height differences in the underlying glacial

bed, or rock. Since these height differences are static, the depressions and the lakes formed in them are always in the same place.

Here, only traces of land ice remain. In this case, the traces have been left by the meltwater that has scoured out a deep trough, close to the disappearing ice near Ny Alesund on Spitsbergen.

If we examine a glacier or icecap from above, we can see a whole pattern of meltwater rivers and lakes. Looking at the snout of a glacier, however, we actually never see a river or stream tumbling over the edge of the glacier. This is because all the meltwater disappears into or under the ice before it reaches the edge of the ice. We can follow the running water over the ice,

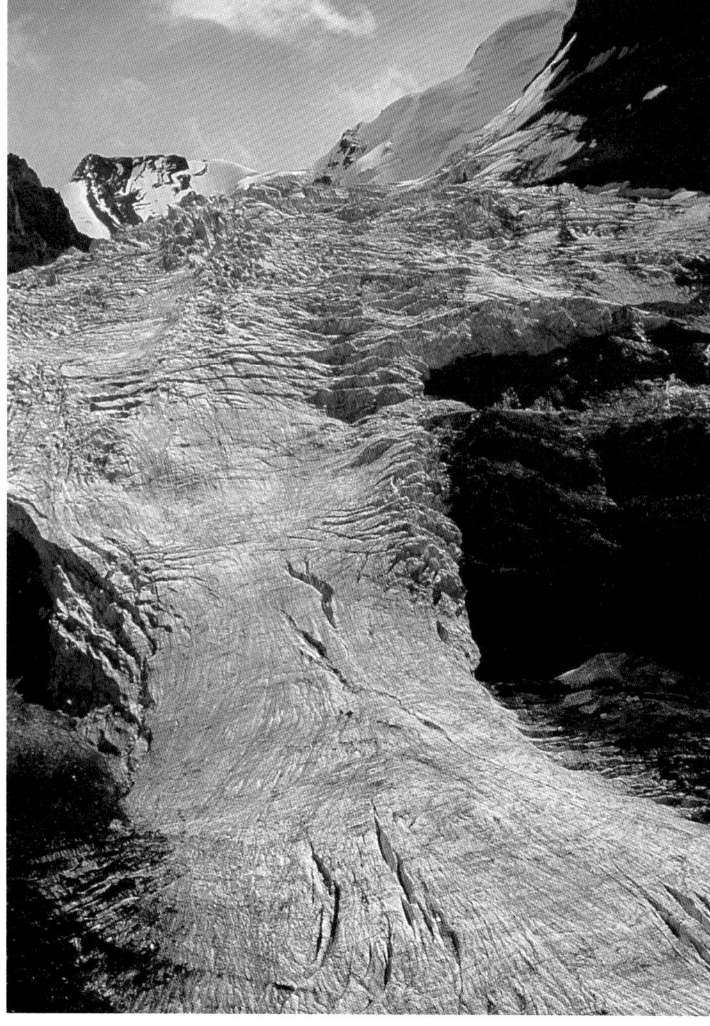

△The glacier ends abruptly in a lake, therefore more ice disappears through calving than through ordinary melting.

◁A glacier flows from a steep slope onto a level valley. It thus spreads out at the foot of the slope in a wide fan. This can also be seen in the longitudinal crevasses in the middle of the glacier.

In the winter the snowline extends over the snout of the glacier. A search for the snowline in this photograph quickly reveals that the Gries Glacier in Switzerland now ends in a reservoir.

The crevasses in the steep section just before the snout of the glacier determine the maximum size of the future icebergs in this region of Antarctica.

The waters around the coasts of Antarctica are strewn with icefloes from broken pack ice and with icebergs from the glaciers that end up in the sea.

The mountains of Patagonia are in many respects terrifying: icebergs break off from the steep ice front of the Perito Moren Glacier, whilst in the background the wind howls over the snow-covered surface.

but as soon as it reaches a crevasse, it disappears down into it, or the water vanishes into a 'moulin', a vertical shaft carved out by the water. Moulins are often the remnants of a glacial crevasse. Such holes also occur on the bottom of lakes on the ice, which we can see when the lake has drained. As we have seen before, crevasses are usually closed up again due to the relentless motion of the glacier, but when a meltwater stream runs into the crevasse, the hole is kept open and even widened by the running water. This hole will only freeze over in winter, although it remains visible because the winter ice in the moulin is different to the normal glacier ice. The water in the moulin freezes from the sides and this produces a structure in the ice that is like the spokes of a wheel. These strange structures are called 'crystal quirks'.

If the water has disappeared under the glacier's surface, it can travel in a number of directions. First, there may be very small openings between the ice crystals that make up the glacier. If the meltwater reaches the bottom of the crevasse, then it can flow through these tiny openings. Another possibility is that a whole network of larger and smaller tunnels exists within the glacier. Investigators of moulins on the Greenland Ice Sheet have discovered giant tunnels and caves, comparable to caves in limestone

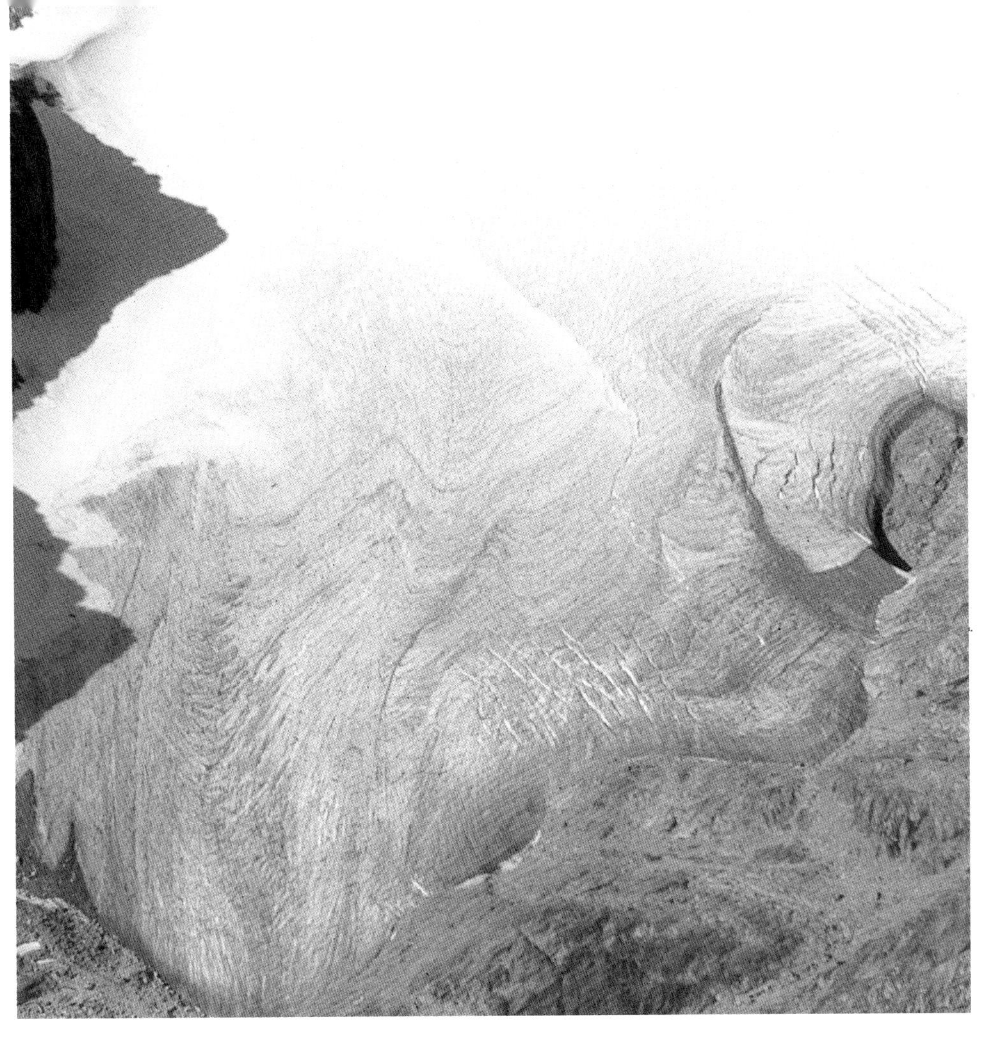

A smooth glacier front in Evigheds Fjord, Greenland. The internal, layered structure can be seen easily because of the melting of the glacial ice.

Icebergs are fascinating: their colour changes with the sun, clouds and rain, whilst melting causes their shape to change continuously. Perito Moren, Argentina.

regions. In Sweden and in the Alps too, tunnel systems have been mapped, for example, under the Gorner Glacier in Switzerland. However, the tunnels formed by the meltwater are not very stable: in the winter there is much less meltwater and the ice is able to squeeze the tunnels closed. Although only ever completely closed at certain points, the result is that the tunnel system might not be fully open and functional in spring when the snow begins to melt and the meltwater needs to be drained away. After a short time, however, all blockages will reopen and the whole system will

◁▽There is no difference between glaciers that end in the sea or those that end in lakes, such as the Perito Moren Glacier in Argentina. The front is just as steep and in both cases icebergs are formed.

▷*For each metre above water, an iceberg has another eight to nine metres below water. Antarctica.*

In serried ranks the pinnacles created by the formation of crevasses move steadily towards the edge of the ice, waiting for their turn to drift away as icebergs from Spitsbergen.

work, as in previous years. The presence of properly functioning tunnel systems can have surprising consequences. During summer thunderstorms, a great deal of water can fall on the glacier. This water quickly collects and drains into the meltwater rivers within a few minutes of the rain starting, causing these rivers to rise quickly. If there were no tunnels at all, the water would be held for some time in and on the glacier and would only gradually reach the meltwater river at the snout of the glacier.

Occasionally we are afforded a brief view of the water network in a glacier. Such is the case when the roof of a tunnel collapses, for then we can see the water flowing deep in the ice. It may also be that the glacier ice has to turn a bend, whilst the water cuts the corner, and then we see a waterfall emerging from under the ice. At the bottom of the steep drop, the water immediately disappears back under the glacier.

Where many glaciers empty into a lake or where glaciers move rapidly, there may be many icebergs. This is the famous lake of Jökullsarlon in Iceland.

Icebergs blown close to the shoreline by the wind often run aground in the shallow coastal waters. If the wind changes direction then they set off again.
In the background of this photograph from the Jasper National Park in Canada, we can see a cross section of a medial moraine, where a thin layer of debris protects a ridge of ice.

Closer to the snout, all the meltwater gathers into one or several large tunnels. The tunnel system in a glacier is just like a large tree: the trunk begins at the snout of the glacier and then splits into ever smaller branches, these being the small tunnels and crevasses. The large tunnels at the snout are often visible on the surface of the glacier because they collapse, forming an elongated hollow.

All the meltwater produced in and on the glacier finally ends up at the snout. In the spring when it starts melting, we can see the water appearing from under the edge of the glacier, and during the course of the summer as the tunnel system develops, the point where the water appears, changes into a large cave opening, the glacier gate. Despite the fact that such a gate seems to stand open invitingly, entering one is

△▷Many ships do not have too many problems with icefloes from pack ice: sailing slowly they push them aside. It is a different story, however, when a large iceberg looms between the icefloes. Antarctica.

certainly not without danger. Pieces of the roof of the tunnel can collapse and the level of the meltwater river can rise suddenly without warning, especially after a sudden shower of rain. In addition to the glacier gate there are other places where meltwater can appear. If we walk along the glacier, water can be seen emerging from crevasses, whilst sometimes it even seems to be appearing out of the glacier's side. In this latter instance, it is clear that the meltwater makes use of the small spaces between the ice crystals.

Glaciers and volcanoes

On 15 November 1985 the world was shocked by the news of a huge disaster in Colombia, South America. An eruption of the glaciated volcano Nevado del Ruiz caused a huge mud flow, wiping out the town of Armero and several small villages. The aftermath of this was a large number of dead and enormous damage to farms and buildings.

In many places on earth there are glaciers on volcanoes; we have already seen this in the first chapter, where different types of glacier are mentioned. This combination of ice and fire is particularly dangerous, because the ice can melt as a result of a volcanic eruption. Huge amounts of water are suddenly released, and on steep slopes, this always leads to the formation of rapidly moving mud flows. Such mud flows can be of immense volume and are also known as 'lahars'.

Several different processes can be distinguished in respect of volcanism under glaciers. The first is the situation in which only heat escapes to the surface over a long period of time (i.e. through 'solfataras'), melting the ice lying on the volcano. This situation, for instance, arises with the Boulder Glacier on Mount Baker in the north-west of the United States, where, following heavy rain, relatively small mud flows regularly flow over the surface of the glacier.

When the water flows down through the glacier, it can make the glacier itself unstable, so that it begins to move more rapidly or even to break up, with all the consequences of such an event. There are also examples of this in South America. If there is a depression in the bed under the ice, then the water can be temporarily collected and held in a sub-glacial lake. The best known example of a lake developing under a glacier is the Grimsvötn in Iceland. The meltwater released from the ice collects here, in a large volcanic depression (caldera) under the icecap Vatnajökull, until it reaches

A small piece of glacier ice acquires the characteristic appearance of shells from melting in the water.

a certain level. The water then overflows from under the icecap, resulting in water flooding out over the outwash plain that spreads from the icecap to the sea. On average, this occurs once every five years. However, when a large volcanic eruption occurs under a glacier in high mountains, such as the Andes, this can melt the glacier away outright, creating a huge flood of

This large iceberg in Antarctica not only contains the remnants of a meltwater tunnel, but also shows the shadowed edges of ledges created when the iceberg lay a different way in the water.

water that picks up rock fragments, weathered materials and vegetation, creating a mud flow. Even if the ice is not all melted at once, the majority of the remaining ice will melt through the heat caused by friction. It is not even necessary for the volcano to lie directly under the glacier, for the accompanying earthquakes can cause an avalanche of ice that has the same effect. This was the case, for example, in 1970 in Peru, where a large ice avalanche was caused by an earthquake (thus without volcanic eruption) on the Huascaran Glacier. As it spilled down the mountain much of the ice melted, resulting in a mud flow. As far as the inhabitants of the town were concerned the effect was the same as for those affected by the Nevado del Ruiz; the town of Yungai was completely destroyed and more than 30,000 people died.

In the late summer of 1996, the world's attention was captured by the eruption of the volcano Bardarbunga under the Vatnajökull icecap in Iceland. Spectacular images were transmitted all over the world, together with stories about the dangers posed by floods. What made this eruption so special was that under this icecap, the largest glacial mass in Europe, lies a caldera that acts as a catchment basin for meltwater, the Grimsvötn caldera as mentioned above. This had emptied in 1995 and was able to collect much of the meltwater that formed. When the lake overflowed a month later it contained about three cubic kilometres of water, the vast majority of which flowed into the sea in a single day. This flood wave destroyed several bridges on the only road around Iceland, isolating the inhabitants of the south-eastern part of the country, and forcing a detour of up to 1,000

kilometres to reach Reykjavik. Although three cubic kilometres of water is a large quantity, it is certainly not a record. When the volcano Katla erupted in 1918 under another Icelandic icecap Myrdallsjökull, it produced at least eight cubic kilometres of water. More than half of this water flooded away in less than eight hours, at such a high speed that a person running ahead would have been overwhelmed by the flood waters. Luckily the coastal plains of Iceland are not as densely populated as some valleys in the Andes, and there was only material damage.

Most glaciers that end in the sea, form relatively small icebergs because the size is determined by the crevasses. Only the great ice plateaus along parts of Antarctica can calve icebergs of several hundred square kilometres.

Glaciation in the past

The geological history of the earth has known several, very long, cold periods. Whilst the warmer periods of the earth's history are known as 'greenhouse' periods, the colder periods are described as 'ice house' periods. Each of these ice house periods consists of a number of cycles

◁A gap in an iceberg does not always have to be the remnants of a tunnel. If ice breaks away from the iceberg on two sides, then a gap can also form.

In clear weather, icebergs do not really constitute a danger to shipping. At night and in bad weather the situation is different; either they cannot be seen or they move unpredictably because of storms, wave action and currents. Fortunately there is not much shipping on the Jökullsarlon lake in Iceland!

In contrast to most glaciers and icecaps, icebergs and icefloes provide a much more welcoming environment for animals. Icebergs and icefloes in the Arctic region are commonly the territory of polar bears.

Seals are the favourite prey of polar bears. In Antarctica, this Weddel seal does not have to face that danger, but they can fall prey to leopard seals and killer whales.

▷A large iceberg off the coast of Antarctica has a basin filled mostly with fresh meltwater.

of warm to cold and back again. The coldest parts of the cycle are known as 'ice ages'. Thus a cold period of more than 100 million years has a number of ice ages or glacials. In many places on earth we can find traces that confirm the existence of these ancient ice ages, usually in the form of unusual types of rocks. In geological terms, the earth is at present in a cold period. The last ice age only ended 10,000 years ago. During that ice age, large sections of the northern hemisphere were covered in ice. The icecap that covered North America was larger than that of present-day Antarctica. Naturally such icecaps leave behind traces of their presence.

Although these large icecaps have disappeared again, this does not mean that existing glaciers are not changing. On the contrary, glaciers constitute part of the climatic system on earth, and if the climate changes, then glaciers will also change. This is quite easily seen in the 'Little Ice Age' that lasted from the middle of the 17th century to the beginning of the 20th century. Paintings from earlier times and photos from the latter days of that period clearly show that glaciers were larger then. Since the beginning of the 20th century, glaciers have been receding rapidly, and this withdrawal has been mapped. From these maps, we know precisely which parts of the land were covered with ice 100 years ago, and even if we did not have these

Waiting for the next storm, a group of stranded icebergs lies along the shoreline of the lake Jökullsarlon.

maps, we would still know how large the glaciers were, for example, because the plants are still adapting to the change. Around the snout of glaciers in high mountains like the Andes, we often see a carpet of vegetation that is different to the vegetation in the wider locality. The former large size of the Obere Grindelwald Glacier in Switzerland, can, for instance, be seen where willows replace coniferous trees.

Glaciers, however, do more to shape the landscape. If the top of a mountain is covered on various sides by glaciers, then these glaciers will gnaw away at that mountain. In the course of time, so much will be eroded away that a steep-sided mountain peak, called a horn, is all that is left. The best known example of this is, undoubtedly, the Matterhorn in Switzerland, but similar mountains can be found in all high mountain ranges, all over the world. Thus glaciers leave their mark on the grand scale of whole mountains, but they also leave behind all kinds of traces of their passage on a much smaller scale.
A glacier that slides downhill, passes over many sorts of rocks. On the upstream

The different structures in glacier ice cause icebergs to melt away irregularly, but also create the most amazing, rugged shapes. Antarctica.

(uphill) side the ice will scrape and polish the rocks till they are smooth, but on the other side, the downstream or downhill side, the glacier will pluck away whole angular blocks of rock. The final product of these two processes is an elongated rock, smoothed and rounded on one side, rough and steep on the other. Such rocks are known as rock knobs or more commonly by the French term 'roches moutonnées'. This name, meaning 'fleecy rocks', is due to their resemblance from a distance to sheep lying in the field.

The smoothed side of the rock knob has a large number of striae, small scratches caused by stones in the ice. If a stone that is under the ice is scraped over a rock, then that stone will scratch the rock and the scratch will reveal the direction in which the ice was moving. Sometimes such a stone can make other shapes. In particular, if the stone is pressed so hard against the rock that it cannot move with the ice, then great pressure builds up between the stone and the rock. If that pressure builds up so high that the stone suddenly moves in a single rapid motion, a semi-circular fragment will break out of the rock. Such a half-moon fragment can have a diameter of nearly one metre. Usually the stone is caught on the rock again and the same mark forms again. We then regularly come across series of such half-moon shaped gouges. Just like the striae, they, too, betray the direction of movement of the ice, even if the ice disappeared thousands of years ago.

As we have already seen, glaciers transport much debris at various levels in the ice. The debris that is held in the sides is deposited in lateral moraines, whilst at the snout it piles up in a terminal moraine. These moraines survive the climatic changes that cause the glaciers to disappear. This is how it is possible to see all sorts of landforms near an active glacier, whilst further away from the glacier, similar forms are present that date from the periods in which the glacier was larger. In many mountain chains we can see, for example, a whole series of lateral and terminal moraines many kilometres distant from the glacier. Most of them have formed in the last 10,000 years, since the end of the last ice age.

In all parts of the world that were covered in glaciers and icecaps during the last ice age, moraines can be found where ice no longer exists. In many cases, large icecaps form large moraines, because in part they act like bulldozers. In this instance, huge plates of rock or sediment are piled up into push moraines. Such terminal and recessional moraines are found in a wide area of north-western Europe in a belt that runs from the Netherlands through Germany and Poland. Similar landforms also occur in the United States and Canada. As glaciers in high mountains are much smaller, the moraines they leave behind are also smaller.

Often the direct drainage of meltwater is hampered by the formation of moraines, therefore there are several strange forms of meltwater channels to be found. The turbulent meltwater can, for instance, scour out a deep cleft in a mountain slope. This cleft follows the direction of the glacier and not that of the slope like a normal river. When the glacier later melts even more, then a new cleft can be scoured out on a lower part of the slope. The first cleft to form remains on the slope as a dry riverbed. There are many other traces associated

with the former presence of glaciers and icecaps, for instance, extensive deposits of ground moraines. These are not deposits that leave behind spectacular scenery, but in the northern hemisphere they form the basis for much fertile agricultural land. In contrast, there are other regions where rocks scraped entirely bare carry the traces of the land ice, for example, in Finland or Canada, whilst the most spectacular and majestic mountain landscapes are also due to the actions of glaciers and not to that of running water.

Would the beginning of a new ice age look like this?

There are many ways to enjoy the beauty of glaciers. On the one hand, there is the spectacular, and sometimes, daunting splendour of existing glaciers and ice masses, and on the other hand, the beauty of the traces left behind in the landscape by the land ice when it was much more extensive than now.